ALL ABOUT ANIMALS

HORSES

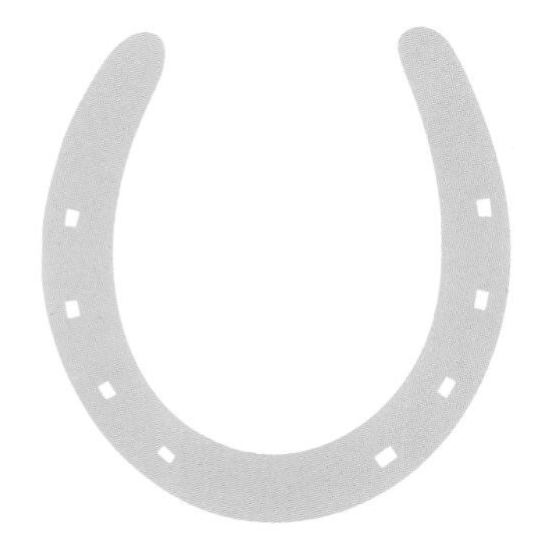

CHELSEA HOUSE PUBLISHERS
A Haights Cross Communications Company ®
Philadelphia

First hardcover library edition published
in the United States of America in
2005 by Chelsea House Publishers,
a subsidiary of Haights Cross Communications.

A Haights Cross Communications Company ®

www.chelseahouse.com

Library of Congress Cataloging-in-Publication Data
ISBN: 0-7910-8685-2

The series *All About Animals* was created and produced by:

McRae Books Srl
Borgo S. Croce, 8, 50122, Florence (Italy)
info@mcraebooks.com
e-mail: mcrae@tin.it
www.mcraebooks.com

Publishers: Anne McRae and Marco Nardi
Text: Rupert Matthews
Illustrations: Fiammetta Dogi, Paula Holguín, Sabrina Marconi, Studio Stalio (Alessandro Cantucci, Fabiano Fabbrucci, Margherita Salvadori)
Graphic Design: Marco Nardi
Layouts: Rebecca Milner, Laura Ottina
Picture Research: Helen Farrell, Chris Hawkes, Claire Moore
Editing: Chris Hawkes, Claire Moore
Color separations: Litocolor, Florence (Italy)
Cutouts: Alman Graphic Design

Printed and bound in China

Acknowledgments
All efforts have been made to obtain and provide compensation for the copyright to the photos and artworks in this book in accordance with legal provisions. Persons who may nevertheless still have claims are requested to contact the publishers.

t = top; tl = top left; tr = top right; tc = top center; c = center; cl = center left; cr = center right; b = bottom; bl = bottom left; br = bottom right; bc = bottom center

The publishers would like to thank the following sources for their kind permission to reproduce the photos in this book:

Contrasto/Corbis: 16cl, 16b, 17, 21cr, 21bl, 21br, 23bl, 24bl, 30tl, 30bl, 30tr, 34tl, 37b; Farabola Foto (Bridgeman Art Library): 32t; *Guidoriccio da Fogliano all'assedio di Monte Massi*, Siena, Palazzo Pubblico (Scala Group, Florence): 27tr; Lonely Planet Images: John Hay/LPI 23tr, David Tipling/LPI 25; The Image Works: ©Eastcott-Momatiuk/The Image Works 9cr, 14b, ©Tom Brakefield/The Image Works 15br, ©Topham/The Image Works 24tr, 28bl, 36tr, ©David M. Jennings/The Image Works 29t, ©Syracuse Newspapers/Al Campanie/The Image Works 31tr, ©Syracuse Newspapers/Mike Greenlar/The Image Works 36tl; Redrawn from 'Black Beauty', illustrated by John Berry, copyright © Ladybird Books Ltd, 1986: 33tl.

p. 14: extract from 'Discoveries: Illiteratus Princeps' by Ben Jonson (1640); p. 25: Ian Fleming (1908–1960) quoted in The Sunday Times, London, October 9th, 1966; p. 27: quotation by Benjamin Disraeli (1804–1881); p. 30: quotation by Abraham Lincoln (1809–1865); p. 33: extract from 'Apocalypse', ch. 10, by D. H. Lawrence (1931); p. 37: extract from 'Guesses at Truth' by Augustus Hare (1827).

ALL ABOUT ANIMALS

HORSES

Rupert Matthews

Contents

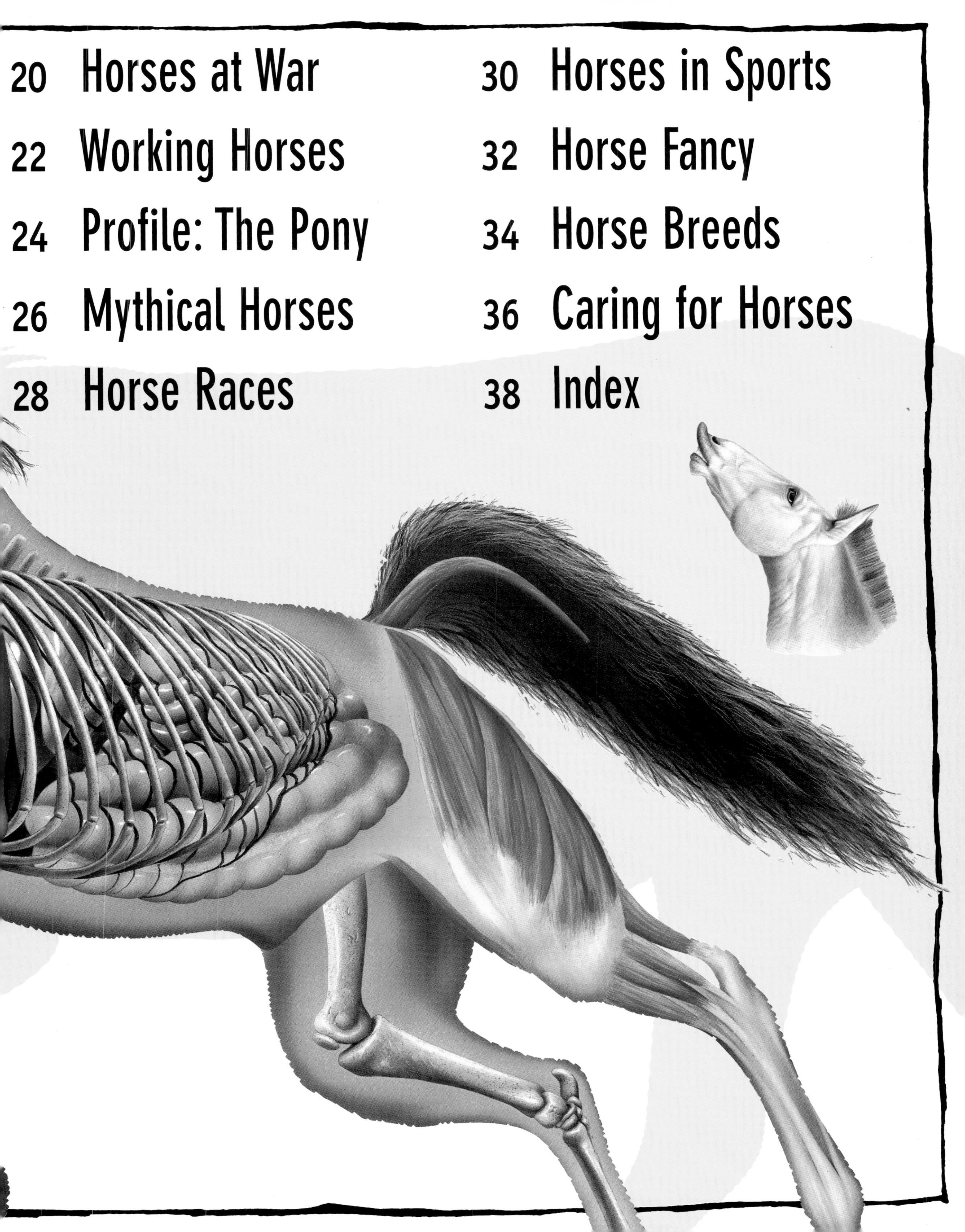

What Makes a Horse a Horse?

Horses are elegant, graceful animals filled with spirit and fire. People have prized the characteristics of these noble creatures for centuries, particularly ancient civilizations, who valued their strength and speed — features that remain an essential part of the horse's appeal today. A horse's loyalty and strength are qualities that can be matched by no other animal. Even though other modes of transport may have replaced the horse in modern times, these animals still retain their magic and attraction.

This life-sized statue of a racehorse (below), complete with a boy rider, was made in Greece over 2,000 years ago.

➲ Speed

Wild horses had to escape from predators, such as wolves or lions, and as a result evolved with the ability to run quickly. Some types of horse are bred specifically for speed and are able to reach about 40 miles per hour (65 km/hr) over short distances. These horses are used today in sports such as horse racing.

A horse gallops with its lungs, perseveres with its heart, and wins with its character.
Anonymous

➲ Sure-footedness

Horses are able to climb steep hills and walk along narrow tracks. They can be used to carry people through mountainous areas where there are few wide roads or paths. This makes them a far more efficient means of transport than cars or trucks in the remote areas of the world.

Loyalty

Horses are social animals that prefer to live in groups. They are renowned for the loyalty and affection that they display toward their human owners and can behave very differently if they are approached by a stranger. Some horses will only allow certain people to ride them.

Agility

The need for wild horses to escape from predators gave them not only speed but also the ability to clear obstacles such as hedges or ravines. Today, sporting events, such as the Badminton Horse Trials in England, test these abilities. Horses and riders compete to see who can finish a course of fences, streams and other obstacles in the quickest time. This French magazine cover from 1905 shows a horse taking part in such an event.

Nobility

The horse projects an image of nobility and grandeur. Throughout history, many kings and important soldiers have been shown in paintings and statues riding a horse — the fact that they are sitting on a horse was thought to add to their importance. This painting (left) shows the soldier Guidoriccio da Fogliano, who led the army of Siena in Italy over 500 years ago.

This statue in Paris, France (below), shows a man taming a wild horse.

Wild and tamed horses

Through time most of the world's wild horses have been captured and tamed by humans. Nowadays, truly wild horses can only be found in remote parts of central Asia. However, in some places, tame horses have escaped into the wild and their descendants now live like wild horses. These horses (shown right) are tame horses gone wild in North America, where they are known as mustangs.

Instinct

Although horses have been domesticated for thousands of years, they retain the instincts of their wild ancestors. In warm, dry weather horses love to roll on their backs in dust. This helps keep the horse's coat free of insects and other pests. Horses also tend to follow each other as if in a herd — an instinct that can be seen in horse races. If a jockey falls off, his horse will continue to follow the others.

Strength and stamina

Until recently, horses were the main source of power for transport. Their strength and stamina enabled them to pull heavy carts and coaches over long distances. Some horses were bred specifically to be able to move heavy loads. They tended to be fairly slow compared to other horses and much larger. Some of these horses are still kept for work in specialist areas.

The Horse Family

The horse family is made up of many different animals, but they all have certain features in common. All horses have a single hoof on each leg and they are all herbivores. Although there are various breeds of horse, they are grouped together into just eight species. The horse family belongs to the much larger group of mammals known as odd-toed ungulates, which also includes rhinoceroses and tapirs.

Przewalski's horse

The wild Asian horse once roamed across the grasslands of central Asia, but by the 19th century it could only be found in Mongolia. A Russian army officer named Nicolai Przewalski discovered the horse when exploring Mongolia in the 1870s. Very few of these animals remain in the wild today, although some herds still live in a semi-wild state in wildlife parks in several European countries. This wild horse is thought to be the ancestor of most of the Asian horse breeds. It has been crossbred with the Tarpan to produce other breeds.

Grevy's zebra

The largest of the wild horses is the Grevy's zebra, which grows to about 4 feet, 11 inches (1.5 m) tall at the shoulder. It lives in small herds on the dry grasslands of East Africa. The animal's coat is marked by many narrow stripes that cover the whole body, except for the belly, which is white. Zebra foals are guarded by an adult male while their mothers feed during the first weeks of their lives.

Plains zebra

Vast herds of Plains zebra, sometimes numbering more than 20,000, form on the grasslands of East Africa. More usually, however, these animals live in family groups of about eight to ten individuals. They graze on grass in woodlands and scrublands as well as on the open plains and rarely stray very far from a source of fresh water.

Asiatic ass

Found in the hills and mountains of Iran, Afghanistan and Uzbekistan, the Asiatic ass is sometimes called the Onager. It climbs into the mountains in the summer to feed on the lush summer grass, and returns to the valleys in the winter to shelter from the cold mountain weather.

Mountain zebra

The Mountain zebra is the smallest — it stands at just 4 feet (1.2 m) at the shoulder — and the rarest of the zebras. At one time they numbered fewer than 1,000, though today there are many more. It was once found across the mountain ranges of southern Africa, but is now only found in a few nature reserves.

Eohippus
14 inches (35 cm) high

Horses grew taller as they evolved.

Mesohippus
18 inches (45 cm) high

Miohippus
21 inches (60 cm) high or over

Evolution

The earliest known form of horse was the *Eohippus*, meaning "horse of the dawn," which lived in North America about 45 million years ago. Also known as *Hyracotherium*, this creature lived in forests. In the millions of years that followed, the descendants of *Eohippus* moved on to open, grassy plains, where they grew larger and developed into fast runners.

Pliohippus
48 inches (120 cm) high

Merychippus
30 inches (90 cm) high

The African ass

Found only in northeastern parts of Africa, the African ass is the wild ancestor of the domestic donkey. It lives in small groups, though there is no firm herd leader or structure, as is the case with most other wild horses. Females often leave the group to have their young, returning a week or two later.

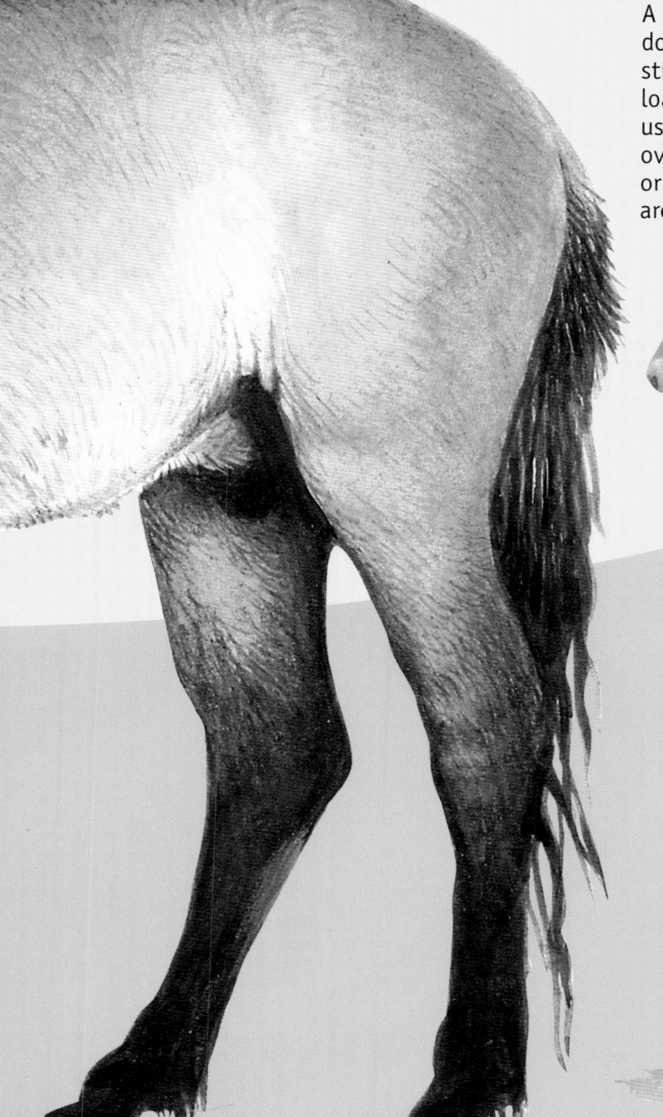

Mule

A mule is a cross between a male donkey and a female horse. It is stronger than a horse and can carry loads for longer periods of time. It is used for carrying people and goods over mountain trails where cars or trucks cannot go. Mules are unable to produce offspring.

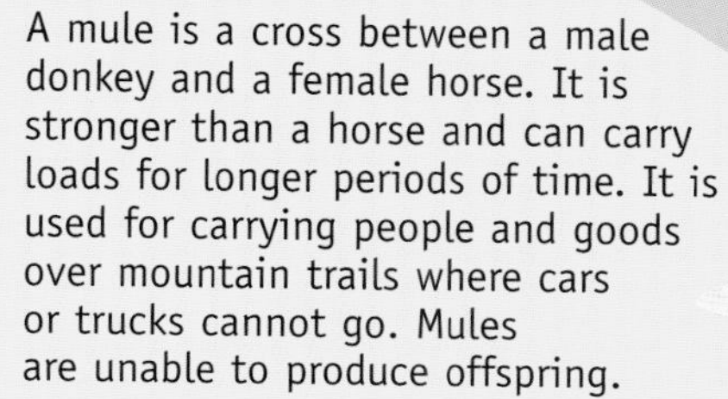

Tarpan

The Tarpan, the original wild horse of eastern Europe, was extinct in the wild by about 1850. It is thought that most of the domestic breeds of horse found in Europe were descended, at least partly, from the Tarpan. During the late 20th century, a special breeding program was carried out to re-create the original wild horse from the domestic breeds of eastern Europe, the closest genetic relations of the original wild animals.

The Horse's Body

The grace, elegance, and beauty of the horse's body are based on an amazingly complex arrangement of bones, muscles, and other organs. Although the many breeds and species of horse can appear to be very different, they all conform to the same basic pattern. The body of the horse is that of a mammal adapted for fast running and for a diet of grass and similar plants. Each part of the body is superbly adapted to its role in supporting the animal as a whole.

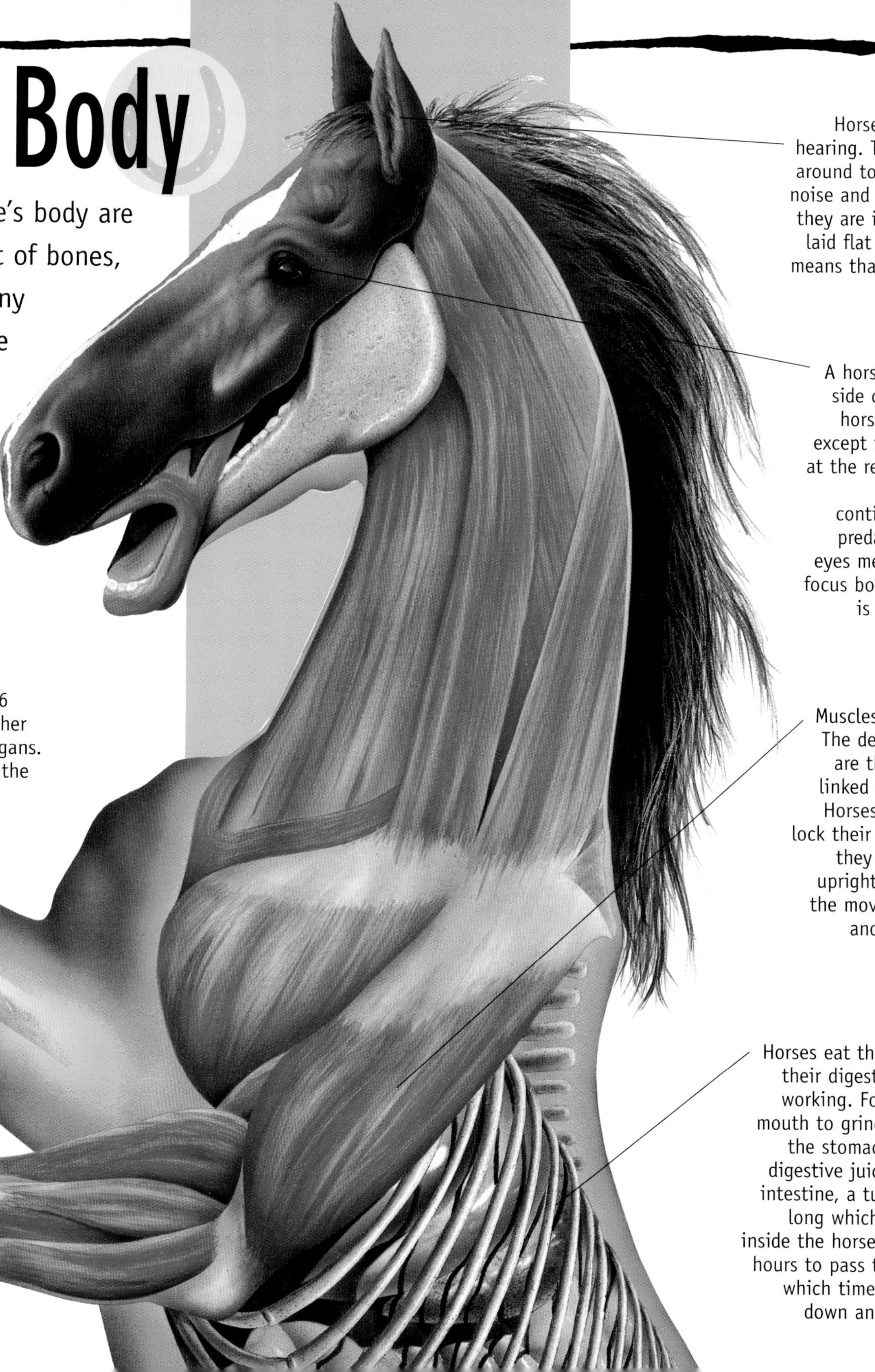

Skeleton

The basic structure of the horse's body is provided by the 216 bones of the skeleton. These connect the muscles to each other and make a protective framework around the body's major organs. The way in which the bones can move against each other at the joints allows the fluid grace of the horse in motion.

Gaits

Wild horses have four basic gaits, or ways of moving their legs. Most tame horses use the same four gaits, but some horses are specially trained to perform extra types of movement that are used in competitions such as dressage.

Ears

Horses have an acute sense of hearing. They can swivel their ears around to focus on the source of a noise and will prick them upright if they are interested. If the ears are laid flat back against the neck, it means that the horse is very angry.

Eyes

A horse's eyes are set on either side of the head. This allows a horse to see all around itself, except for a very small blindspot at the rear. This is useful for wild horses, which need to be continually on the lookout for predators. The position of the eyes means that horses can only focus both eyes on an object if it is directly in front of them.

Muscles

Muscles enable a horse to move. The deep, and wide leg muscles are the most powerful and are linked to the bones by tendons. Horses have a unique ability to lock their leg muscles solid so that they can sleep while standing upright. Smaller muscles control the movements of the head, tail, and other parts of the body.

Digestion

Horses eat throughout the day so that their digestive system is continually working. Food is first chewed in the mouth to grind it up, then it passes to the stomach where it is mixed with digestive juices. It then moves to the intestine, a tube about 98 feet (30 m) long which twists and turns around inside the horse's body. Food takes many hours to pass through this tube, during which time the nutrients are broken down and then absorbed into the animal's bloodstream.

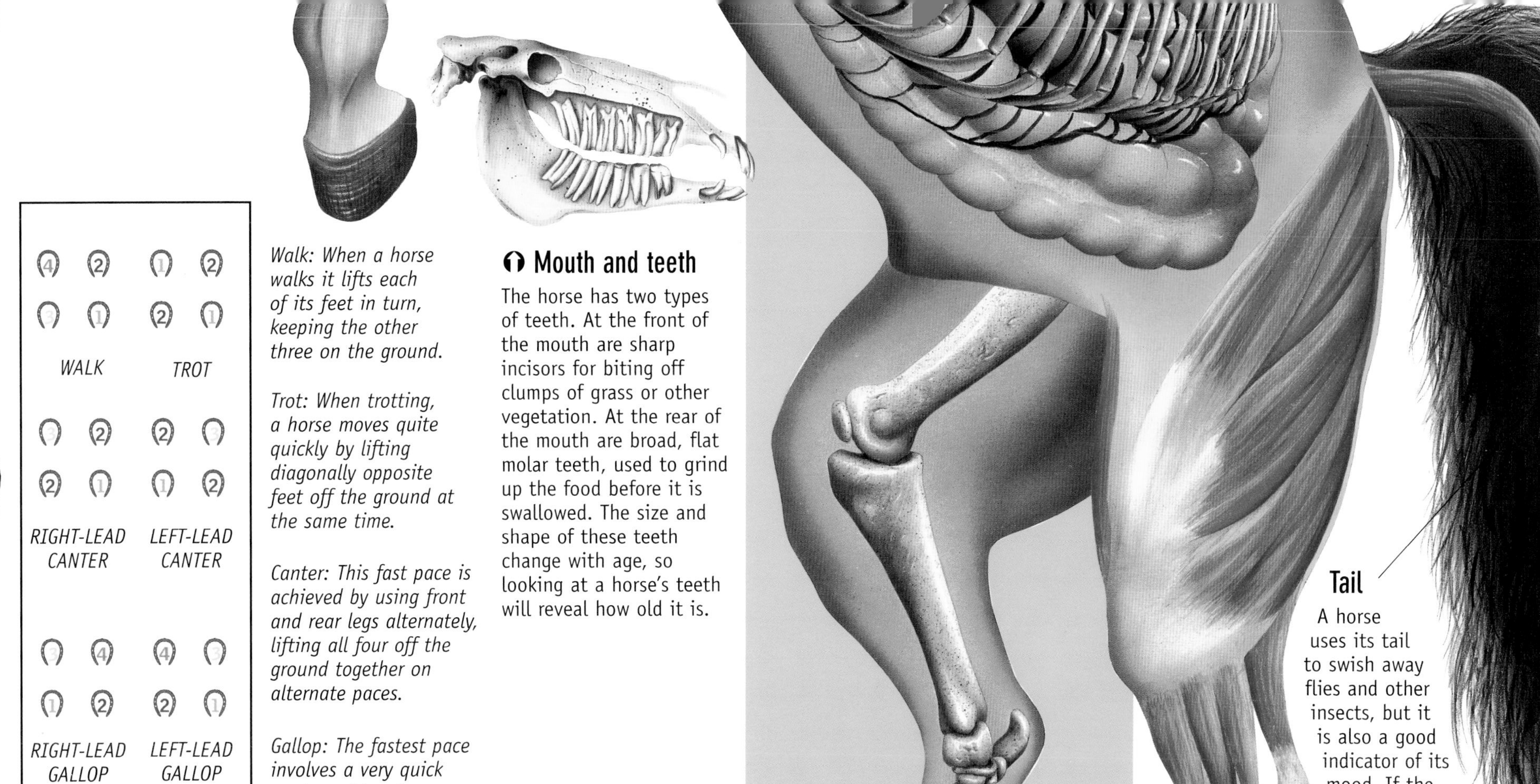

Walk: When a horse walks it lifts each of its feet in turn, keeping the other three on the ground.

Trot: When trotting, a horse moves quite quickly by lifting diagonally opposite feet off the ground at the same time.

Canter: This fast pace is achieved by using front and rear legs alternately, lifting all four off the ground together on alternate paces.

Gallop: The fastest pace involves a very quick canter, with only one or two feet on the ground at any one time.

Mouth and teeth

The horse has two types of teeth. At the front of the mouth are sharp incisors for biting off clumps of grass or other vegetation. At the rear of the mouth are broad, flat molar teeth, used to grind up the food before it is swallowed. The size and shape of these teeth change with age, so looking at a horse's teeth will reveal how old it is.

Tail

A horse uses its tail to swish away flies and other insects, but it is also a good indicator of its mood. If the tail is held upright, it usually means that the horse is interested or alarmed. If the tail lashes quickly from side to side then the horse is angry. If the tail hangs quietly, with a few gentle movements, it shows that the horse is relaxed.

Coat colors

The coats of domestic horses vary widely in colors and patterns. Some breeds have a specific color, but most produce horses of all different shades. The majority have coats of one basic color, but with small markings of other colors. Paints are horses marked by patches of white and one other color. Roans are horses where hairs of different hues are mixed evenly across the body.

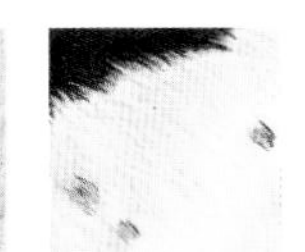
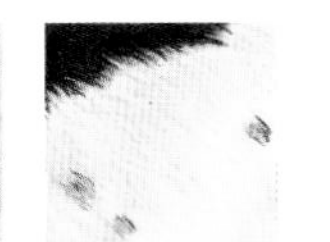

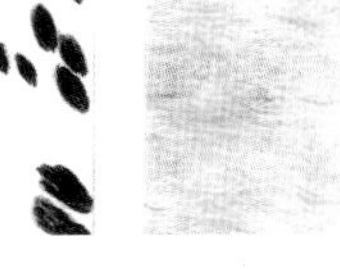
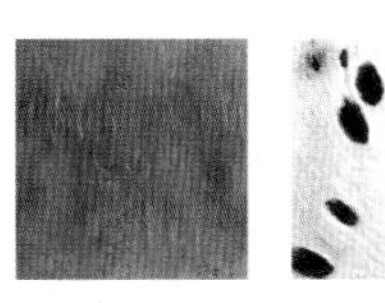

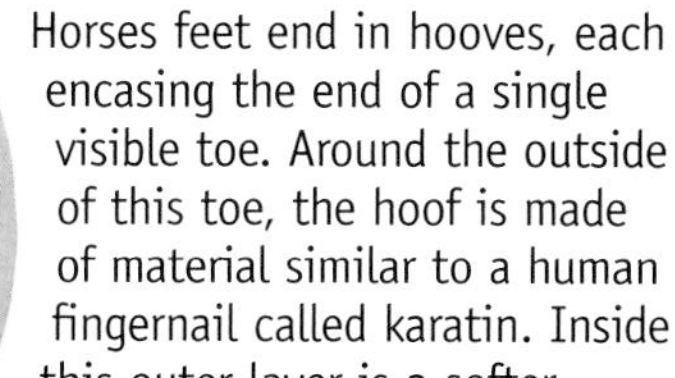

Feet

Horses feet end in hooves, each encasing the end of a single visible toe. Around the outside of this toe, the hoof is made of material similar to a human fingernail called karatin. Inside this outer layer is a softer, spongy center that acts as a shock absorber as the horse moves.

Foals

Horses less than a year old are known as foals. In the wild, foals stay very close to their mothers for at least six months, rarely straying more than a few yards away. Their mothers teach them which plants are good to eat and how to watch out for danger. Domestic foals also like to stay close to their mothers, although they need to get used to humans at an early age if they going to be able to interact with them successfully in later life.

Grazing

A horse's diet is made up almost exclusively of grass. A wild horse will spend many hours each day grazing. Horses generally prefer soft, green grass and may leave an area of dry, brown grass in search of better food elsewhere. When feeding, a horse lifts its head from time to time to look around for any sign of predators or danger.

A Horse's Life

The average life expectancy of a horse is 25 to 30 years, although some may live to be much older. Wild horses live in well defined social structures that helps them survive in hostile conditions. They also have ways of communicating and interacting with each other. Domestic horses behave in similar ways.

They say Princes learn no art truly but the art of horsemanship. The reason is the brave beast is no flatterer. He will throw a Prince as soon as his groom.

Ben Jonson (1573–1637), poet and dramatist

Herds

Wild horses live in herds of up to 30 animals. The lead stallion protects the herd from attack and keeps a watchful eye out for predators. The leading mare also plays an important role within the group. She is responsible for deciding when the herd should go to drink water and where to find the best grass for grazing.

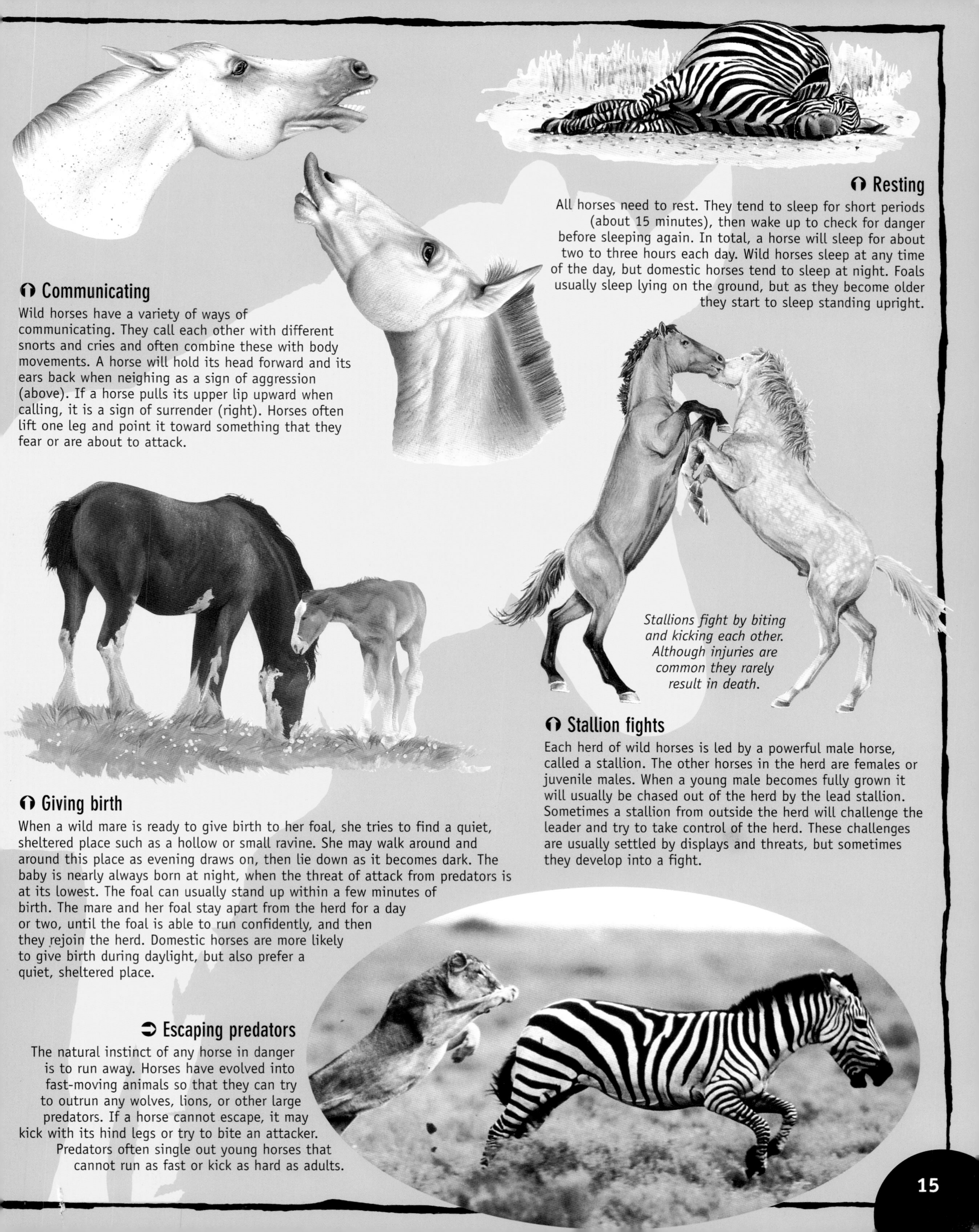

Resting

All horses need to rest. They tend to sleep for short periods (about 15 minutes), then wake up to check for danger before sleeping again. In total, a horse will sleep for about two to three hours each day. Wild horses sleep at any time of the day, but domestic horses tend to sleep at night. Foals usually sleep lying on the ground, but as they become older they start to sleep standing upright.

Communicating

Wild horses have a variety of ways of communicating. They call each other with different snorts and cries and often combine these with body movements. A horse will hold its head forward and its ears back when neighing as a sign of aggression (above). If a horse pulls its upper lip upward when calling, it is a sign of surrender (right). Horses often lift one leg and point it toward something that they fear or are about to attack.

Stallions fight by biting and kicking each other. Although injuries are common they rarely result in death.

Stallion fights

Each herd of wild horses is led by a powerful male horse, called a stallion. The other horses in the herd are females or juvenile males. When a young male becomes fully grown it will usually be chased out of the herd by the lead stallion. Sometimes a stallion from outside the herd will challenge the leader and try to take control of the herd. These challenges are usually settled by displays and threats, but sometimes they develop into a fight.

Giving birth

When a wild mare is ready to give birth to her foal, she tries to find a quiet, sheltered place such as a hollow or small ravine. She may walk around and around this place as evening draws on, then lie down as it becomes dark. The baby is nearly always born at night, when the threat of attack from predators is at its lowest. The foal can usually stand up within a few minutes of birth. The mare and her foal stay apart from the herd for a day or two, until the foal is able to run confidently, and then they rejoin the herd. Domestic horses are more likely to give birth during daylight, but also prefer a quiet, sheltered place.

Escaping predators

The natural instinct of any horse in danger is to run away. Horses have evolved into fast-moving animals so that they can try to outrun any wolves, lions, or other large predators. If a horse cannot escape, it may kick with its hind legs or try to bite an attacker. Predators often single out young horses that cannot run as fast or kick as hard as adults.

Profile: Arabs

The Arabian is the oldest and purest of all horse breeds. Strong, fast, intelligent, and often astonishingly beautiful, it is a favorite around the world. Although the breed is still numerous in Arabia, it is now raised in many other countries too. The Arabian has also been mixed into other breeds to improve their looks and performance. As so many of today's horses have a trace of Arabian ancestry, this breed is often called "the fountainhead of the horse."

Arabian ID

Origin: Arabia and nearby desert regions
Background: bred by desert tribesmen over 1,000 years ago, maybe more
Height: 14 or 15 hands
Colour: usually bay or chestnut, some are grey or black
Uses: riding and racing
Characteristics: speed, endurance, courage, and intelligence

Arab origins

As its name suggests, the Arabian horse originated in the deserts of the Arabian peninsula. It was being bred by Bedouin tribesmen more than 1,000 years ago, and may date back more than twice as long. The religion of Islam made the raising of quality horses a religious duty.

Napoleon riding his famous Arab Marengo.

Emperors' horses

The French Emperor Napoleon (1769–1821) captured a beautiful Arabian stallion when fighting in Egypt in 1799. He named it Marengo and rode it in all his later battles. The horse was captured by the British after Napoleon's defeat at the Battle of Waterloo in 1815 and was used to breed with native English horses.

A Quarter Horse (above), one of many breeds improved by Arab blood.

Body features

The Arab's size, elegant carriage, and physical beauty make it one of the most recognizable of breeds. It usually stands between 14 and 15 hands and is, therefore, rather small for a riding horse. It is surprisingly strong for its size, though, and can carry as heavy a load as some larger animals. Most Arabian horses are bay, gray, or chestnut in color, though a few are black. Unlike all other horse breeds, which have 18 ribs, Arabians have only 17, making their bodies slightly shorter.

The Arabian influence

The characteristics of the Arabian have been bred into other horse types to alter their abilities. The American Quarter Horse was developed to be able to accelerate and run quickly over a short distance (a quarter of a mile). The infusion of Arab blood has given these horses more stamina.

Eyes: These are large and set wide apart on the head and lower down the face than in most other breeds.

Head: The Arab has a small head with a slender muzzle. Its slightly concave or "dished" profile is one of the breed's most recognizable features.

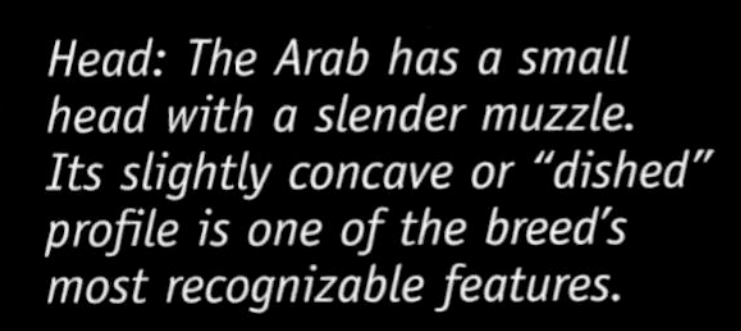

Legs: These are straight and fine with clearly defined tendons and muscles. The forelegs are often rather short below the knee.

Body: This is compact, with well-muscled shoulders and rump. The back often dips slightly in the center.

Tail: Arabians carry their long, fine-haired tails arched and high.

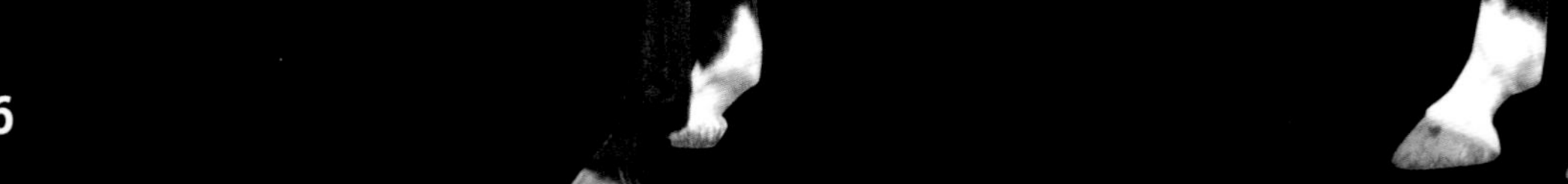

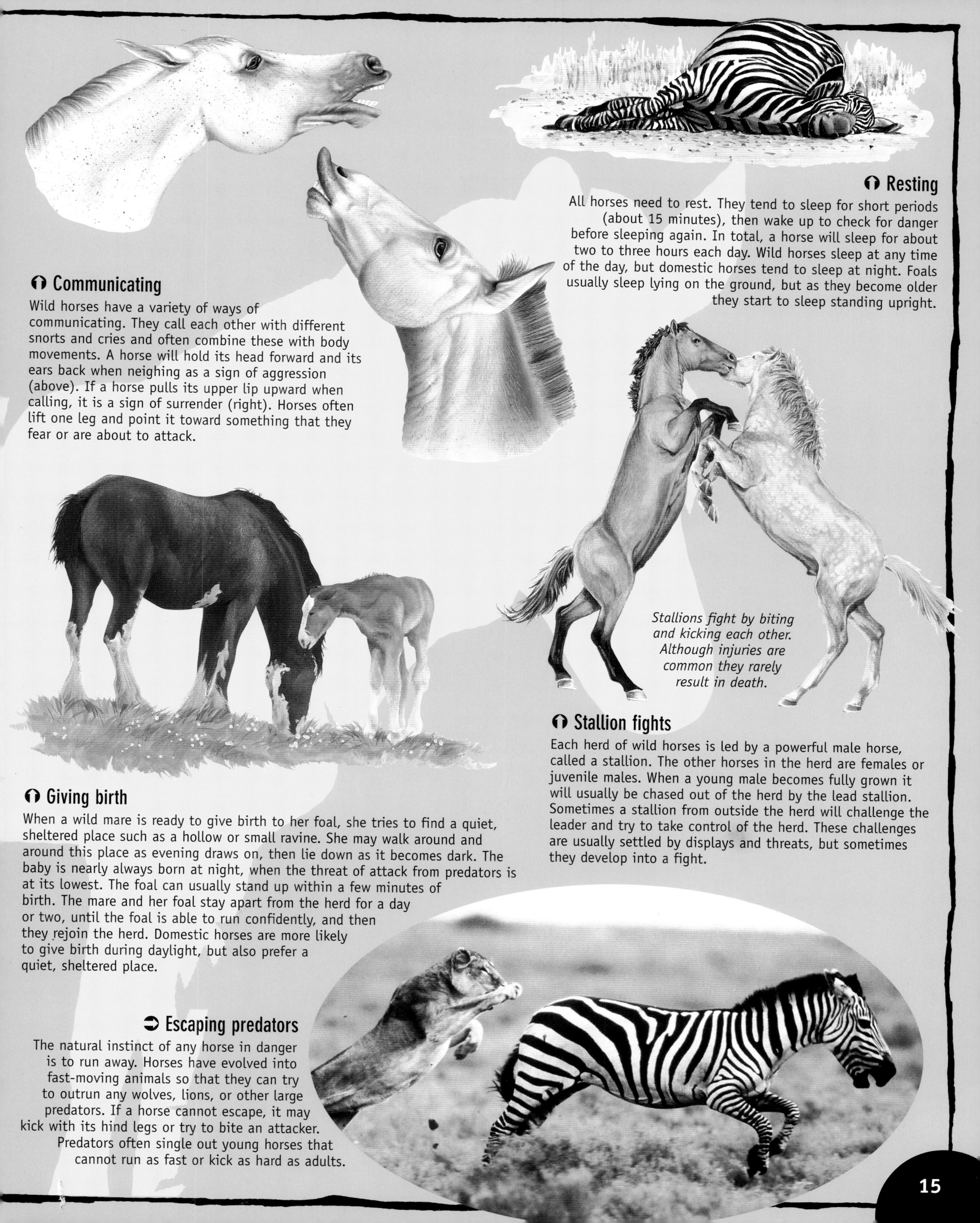

Communicating

Wild horses have a variety of ways of communicating. They call each other with different snorts and cries and often combine these with body movements. A horse will hold its head forward and its ears back when neighing as a sign of aggression (above). If a horse pulls its upper lip upward when calling, it is a sign of surrender (right). Horses often lift one leg and point it toward something that they fear or are about to attack.

Resting

All horses need to rest. They tend to sleep for short periods (about 15 minutes), then wake up to check for danger before sleeping again. In total, a horse will sleep for about two to three hours each day. Wild horses sleep at any time of the day, but domestic horses tend to sleep at night. Foals usually sleep lying on the ground, but as they become older they start to sleep standing upright.

Stallions fight by biting and kicking each other. Although injuries are common they rarely result in death.

Stallion fights

Each herd of wild horses is led by a powerful male horse, called a stallion. The other horses in the herd are females or juvenile males. When a young male becomes fully grown it will usually be chased out of the herd by the lead stallion. Sometimes a stallion from outside the herd will challenge the leader and try to take control of the herd. These challenges are usually settled by displays and threats, but sometimes they develop into a fight.

Giving birth

When a wild mare is ready to give birth to her foal, she tries to find a quiet, sheltered place such as a hollow or small ravine. She may walk around and around this place as evening draws on, then lie down as it becomes dark. The baby is nearly always born at night, when the threat of attack from predators is at its lowest. The foal can usually stand up within a few minutes of birth. The mare and her foal stay apart from the herd for a day or two, until the foal is able to run confidently, and then they rejoin the herd. Domestic horses are more likely to give birth during daylight, but also prefer a quiet, sheltered place.

Escaping predators

The natural instinct of any horse in danger is to run away. Horses have evolved into fast-moving animals so that they can try to outrun any wolves, lions, or other large predators. If a horse cannot escape, it may kick with its hind legs or try to bite an attacker. Predators often single out young horses that cannot run as fast or kick as hard as adults.

Profile: Arabs

The Arabian is the oldest and purest of all horse breeds. Strong, fast, intelligent, and often astonishingly beautiful, it is a favorite around the world. Although the breed is still numerous in Arabia, it is now raised in many other countries too. The Arabian has also been mixed into other breeds to improve their looks and performance. As so many of today's horses have a trace of Arabian ancestry, this breed is often called "the fountainhead of the horse."

Arabian ID

Origin: Arabia and nearby desert regions
Background: bred by desert tribesmen over 1,000 years ago, maybe more
Height: 14 or 15 hands
Colour: usually bay or chestnut, some are grey or black
Uses: riding and racing
Characteristics: speed, endurance, courage, and intelligence

Arab origins

As its name suggests, the Arabian horse originated in the deserts of the Arabian peninsula. It was being bred by Bedouin tribesmen more than 1,000 years ago, and may date back more than twice as long. The religion of Islam made the raising of quality horses a religious duty.

Napoleon riding his famous Arab Marengo.

Emperors' horses

The French Emperor Napoleon (1769–1821) captured a beautiful Arabian stallion when fighting in Egypt in 1799. He named it Marengo and rode it in all his later battles. The horse was captured by the British after Napoleon's defeat at the Battle of Waterloo in 1815 and was used to breed with native English horses.

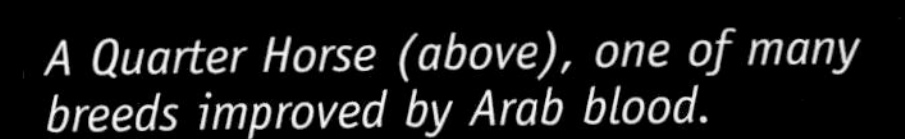

A Quarter Horse (above), one of many breeds improved by Arab blood.

Body features

The Arab's size, elegant carriage, and physical beauty make it one of the most recognizable of breeds. It usually stands between 14 and 15 hands and is, therefore, rather small for a riding horse. It is surprisingly strong for its size, though, and can carry as heavy a load as some larger animals. Most Arabian horses are bay, gray, or chestnut in color, though a few are black. Unlike all other horse breeds, which have 18 ribs, Arabians have only 17, making their bodies slightly shorter.

The Arabian influence

The characteristics of the Arabian have been bred into other horse types to alter their abilities. The American Quarter Horse was developed to be able to accelerate and run quickly over a short distance (a quarter of a mile). The infusion of Arab blood has given these horses more stamina.

Eyes: These are large and set wide apart on the head and lower down the face than in most other breeds.

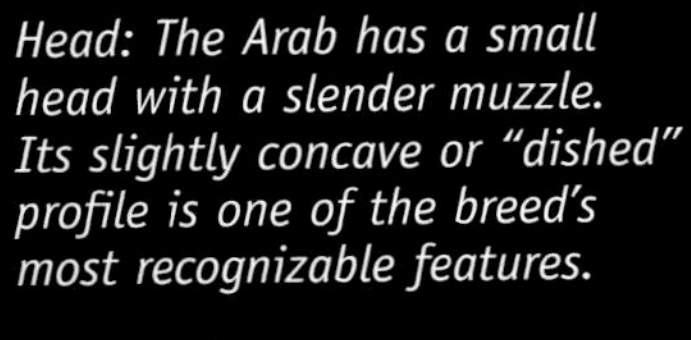

Head: The Arab has a small head with a slender muzzle. Its slightly concave or "dished" profile is one of the breed's most recognizable features.

Body: This is compact, with well-muscled shoulders and rump. The back often dips slightly in the center.

Legs: These are straight and fine with clearly defined tendons and muscles. The forelegs are often rather short below the knee.

Tail: Arabians carry their long, fine-haired tails arched and high.

THE ARABIAN TEMPERAMENT

The Arabian's character is among the most attractive of all horses. It was bred originally to be brave and tough, so that it would not shy away from the noise and dangers of the battles fought by its Arab owners. It was also considered an advantage for the horse to be intelligent, so that it could cope easily with strange situations.

And God took a handful of southerly wind, blew His breath over it and created the horse.

Bedouin legend

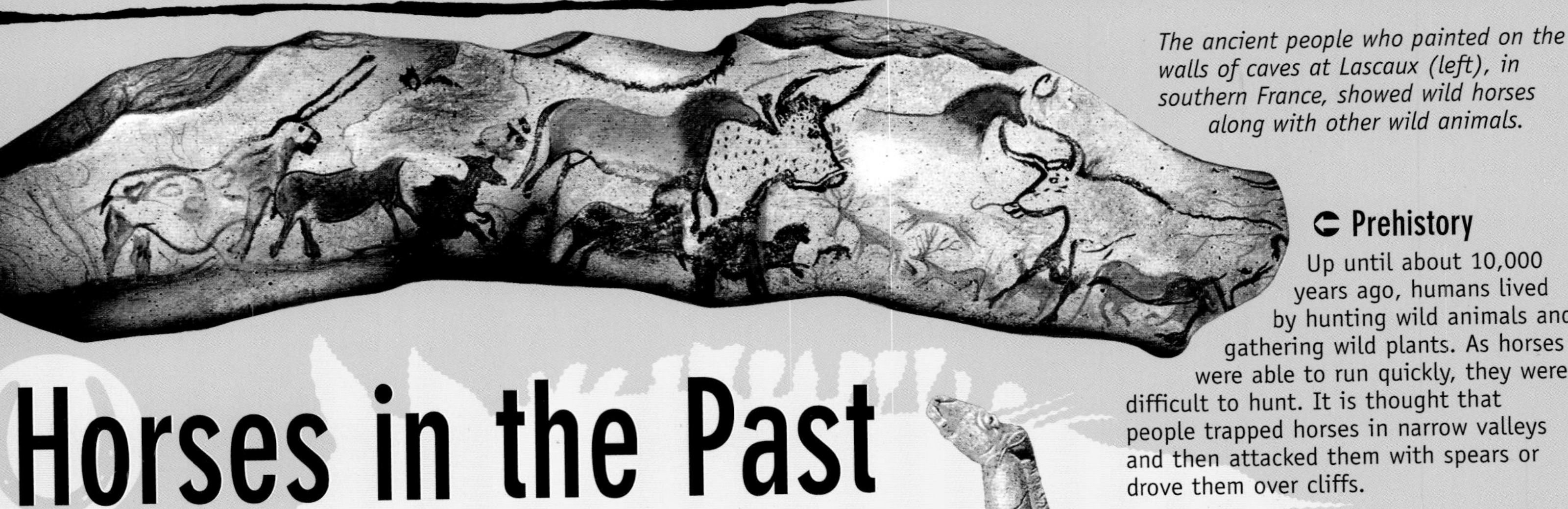

The ancient people who painted on the walls of caves at Lascaux (left), in southern France, showed wild horses along with other wild animals.

Prehistory

Up until about 10,000 years ago, humans lived by hunting wild animals and gathering wild plants. As horses were able to run quickly, they were difficult to hunt. It is thought that people trapped horses in narrow valleys and then attacked them with spears or drove them over cliffs.

This prehistoric spear thrower (left) is decorated with a carved horse.

Horses in the Past

Horses and people have lived in close contact for thousands of years. For much of that time, humans saw horses as just another animal to hunt for food. However, once horses had been tamed, it became clear that they could be used for a wide variety of purposes. Horses became the main source of power for transport and, as such, played a considerable role in the growth of human civilization.

Taming in Eurasia

Scientists believe that the horse was first tamed by humans somewhere on the grasslands of Europe or Asia more than 6,000 years ago. By 5,000 years ago, horses were being used right across Asia and Europe.

Following the reindeer

The horse was not the only animal to be tamed and ridden by humans. The reindeer of the far northern areas of Europe and Asia was tamed about 7,000 years ago. It was used to pull carts and sleighs, but could be ridden as well. Although some reindeer are still used in this way, most civilizations now use the horse instead.

This picture from ancient Egypt shows a donkey carrying grain from the fields to a village.

The first horse peoples

The earliest people to use tamed horses were the tribes who lived on the open plains of central Asia and eastern Europe. Archaeological finds indicate that people were riding horses as early as 6,000 years ago, perhaps using them to control herds of domestic animals or to hunt wild creatures. This felt hanging was found frozen in a 3,000-year-old tomb in Siberia and shows what these early horse people would have looked like.

Domestication of the donkey

The wild ass was tamed about 6,000 years ago in the Middle East. The tame donkeys were not strong enough to carry an adult rider for long periods of time, so they were used to carry packs of goods or to pull carts. The scene below is from the city of Ur and is 4,600 years old. It shows donkeys pulling war chariots into battle.

Draft-horses

Inhabitants of the valleys of the Tigris and Euphrates rivers had been using oxen to pull carts when the first tame horses were brought in from the north about 5,000 years ago. These horses were soon being used to pull carts as they were quicker than oxen. This small model shows a cart being pulled by early horses. By about 4,000 years ago, the first draft-horses had been bred. They were bigger and stronger than riding horses, but were also slower.

The Classical world

After about 600 B.C. new types of saddles and harnesses made riding much safer and more convenient, while horseshoes meant that horses could carry heavier loads without damaging their feet. Chariots began to be used only for parades or sport. The people of ancient Greece and Rome rode horses extensively for travel and war, though they still used draft-horses to pull carts and carriages.

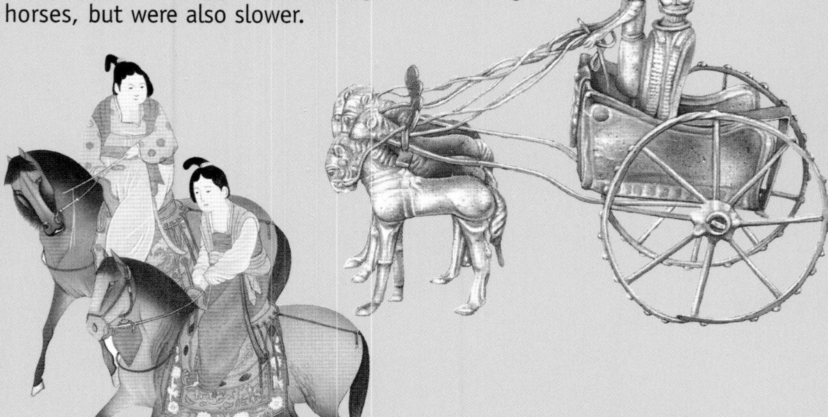

The horse in China

The Chinese developed many important aids to horse riding. One of the most important of these was the stirrup, which appeared about 1,800 years ago. The stirrup allows a rider to sit more securely in the saddle and enables him to ride faster, or to use weapons while riding. These Chinese ladies (above), riding with stirrups, appear in a painting about 1,000 years old.

The New World

Horses became extinct in the Americas about 10,000 years ago. They were re-introduced by Spanish explorers in the late 15th century. At first the Native Americans did not understand that a man on a horse were two different creatures, but thought they were one fantastic animal. Using horses and steel weapons, the Europeans quickly conquered the native peoples. This picture (left) shows mounted Spanish soldiers defeating an army of Aztecs on foot.

Medieval horses

During the period from A.D. 700 to 1500, the battlefield was dominated by men wearing armor and riding horses. These knights practiced fighting each other with blunt weapons. Contests of battle skills, known as tournaments, were held at regular intervals. Jousting involved two knights charging toward each other and trying to knock each other from the saddle using a long pole called a lance.

Horses at War

Horses have been used in warfare for as long as people have known how to ride them. A man on a horse or in a chariot can travel more quickly about the battlefield than a man on foot. The extra speed gives a charging horse more impact than a charging man. These advantages made the combination of man and horse a battle-winning weapon in war for many centuries.

War chariots

Before the development of riding saddles and harnesses, in about 700 B.C., it was difficult to ride a horse and use weapons at the same time. For this reason, horses were used to pull chariots containing armed men. This picture shows the Egyptian pharaoh Rameses II (reigned 1279–1213 B.C.) shooting a bow and arrow while riding in a war chariot.

The Mongols

In 1206, the Mongol tribes of central Asia were united by the ruler Temujin (1165–1207), who used the title Genghis Khan, meaning "great leader." The Mongols fought on horseback and were able to shoot arrows with deadly accuracy while riding at a full gallop. Over the next 50 years, the Mongols conquered a vast empire covering China, Iran, central Asia, and much of eastern Europe.

Christian knights

When Europeans were attacked by the light Muslim cavalry they responded by using heavy cavalry. Knights, wearing sturdy metal armor, were mounted on large, powerful horses. A charge of knights could defeat the lighter horsemen. The knight became the ideal of medieval Christianity.

Muslim warriors

The Muslim religion was founded by the prophet Muhammad (570–632), who raised an army to spread his religion to neighboring peoples. The key part of the army was a large force of lightly armed cavalry that could move quickly to mount surprise attacks on the enemy. Muhammad made breeding quality horses a religious duty for his followers.

A statue of the Spanish knight El Cid on his horse Babieca.

A pair of stirrups made for Genghis Khan.

The new cavalry

By about 1600, firearms were in common use in battle. A cavalry could no longer break the infantry formations because of the increasingly effective use of guns and because the horses tended to shy away from loud noises. Horsemen began to equip themselves with pistols as well as with swords or lances. These men (above) are training a warhorse to ignore threatening behavior and loud noises.

Horses at Waterloo

The Battle of Waterloo in 1815 saw the defeat of the French Emperor Napoleon by the British under the Duke of Wellington (1769–1852) and the Prussians under Marshal Blücher. The battle saw a mass charge by more than 20,000 French cavalry, but the attack failed to defeat the British infantry. The vanquished French were chased from the field by the British and Prussisan cavalry.

A statue of the Duke of Wellington on his horse Copenhagen.

Alexander the Great and Bucephalus

Alexander the Great (356–323 B.C.) of ancient Greece conquered a vast empire by using his heavily armored cavalry to win many battles. Alexander always rode his horse Bucephalus into battle, since it was immensely strong and was specially trained for battlefield maneuvers. Bucephalus was killed at the Battle of Hydaspes in 326 B.C., and, according to historians, Alexander gave the horse a funeral and wept.

The Charge of the Light Brigade

At the Battle of Balaclava in 1854, the British light cavalry brigade attacked the Russian artillery instead of the infantry because of muddled orders. The beautifully trained horses rode at full gallop toward the artillery without flinching or panicking, but were cut down. Of the 673 horses that took part in the charge, more than 500 were killed or were so badly wounded that they had to be shot.

Horses in the World Wars

Cavalries were of no use in the trenches during World War I, though they were used in the deserts of the Middle East. In World War II, cavalries were used to carry men quickly across rough terrain where tanks or trucks could not travel easily.

Custer's last stand

In 1876, the famous American lieutenant colonel George Armstrong Custer (1839–1876) led his regiment, the 7th Cavalry, to attack a large force of Sioux and Cheyenne tribesmen in the American West. Custer was confronted by overwhelming numbers of tribesmen fighting as light cavalry. Every man in Custer's command was killed. The only survivor found by the reinforcements was a wounded horse named Comanche.

Warhorses today

Many modern armies have regiments of cavalry that parade in splendid uniforms during important celebrations, but the regiments fight in tanks or armored cars during wars. However, some armies also have fighting cavalry units for use in mountainous areas or in regions that do not have any roads. Cavalries were used extensively during the Russian-Afghan war in the 1980s. This picture (below) shows the French cavalry during a parade in Paris.

Working Horses

Most horses today are used for riding, for fun or for recreation, but until only recently, horses were used mostly for work. Transportation systems relied on horsepower, and special breeds of horse were developed to suit different types of work. Even today some jobs still require the efforts of a horse to get them done. The working horse is still with us.

Hunting with horses

A person on a horse can move more quickly than a person on foot and can catch fast moving prey. Few people rely on hunting for food nowadays and so most hunting from horseback is carried out for sport. Hunting in this way makes great demands on both horse and rider, so only the most experienced can hunt successfully.

Native Americans hunted bison on horseback.

A horse pulls a seed drill, a device that plants a seed at the correct depth in the soil.

Agricultural horses

In the past, horses were used to operate a variety of farm equipment, including ploughs, drills, harvesters, and mills. As agricultural machines became increasingly sophisticated from 1700 on, horses replaced oxen as the main source of power. A horse is much easier to control, can produce a steadier source of power, and can work more quickly than the slow ox.

It's not the horse that draws the cart, but the oats.
Russian proverb

PONY EXPRESS!
CHANGE OF TIME! REDUCED RATES!
10 Days to San Francisco!
LETTERS
WILL BE RECEIVED AT THE
OFFICE, 84 BROADWAY,
NEW YORK,
Up to 4 P. M. every TUESDAY,
AND
Up to 2½ P. M. every SATURDAY,
Which will be forwarded to connect with the PONY EXPRESS leaving ST. JOSEPH, Missouri,
Every WEDNESDAY and SATURDAY at 11 P. M.
TELEGRAMS

Helping communication

Horses have been used to carry messages for centuries. The Persian Emperor Cyrus started to build a "Royal Road" across his lands in 540 B.C. that only royal messengers were allowed to use. In the United States, the Pony Express linked the east and west coasts in just ten days' hard riding. Horses and riders were changed frequently during the course of the journey.

Herding cattle

Horses used for herding cattle are trained to turn suddenly or to stop abruptly, so that they can be in the right place for whatever task the rider wishes to carry out. Each cowboy has several horses, some trained for specific jobs and others used to replace horses that become tired from the exertions of a day's work.

Cowboys rounding up a herd of cattle at day break (right).

A scene from New York in the 1880s. Horses were used to pull both private carriages and public trams, while a steam engine pulls a train on an elevated railway.

Passenger transport

In the days before cars, only wealthy people could afford their own carriages, so most people paid to ride on a public tram or an omnibus — both of which were pulled by horses. Coaches also ran between cities and could cover hundreds of miles in a single journey.

Circus acts

Trick riding and performing stunts on horseback have been popular entertainments since the days of ancient Greece. The first modern circus was opened in London in 1769 by a retired soldier called Philip Astley. Astley performed many riding tricks himself, and later added clowns and jugglers to the act.

Leaping on or off a bareback horse (left) has been a popular circus trick for centuries.

Parading

Horses have traditionally been used in parades, to carry soldiers or to pull ceremonial coaches. Although motorized transport is now used in everyday life, horses are still used for special events, as they lend dignity and nobility to the procession.

Police horses

Police forces often used mounted officers to cover large areas of open land, such as is found in northern Canada. In more recent years, mounted police have proved a useful tool in controlling crowds at parades or at sporting events. Mounted police may also be used to control riots and other scenes of disorder.

Right: A Canadian Mounted Policeman, or Mountie.

A parade moves through the streets of London (left), carrying Queen Elizabeth II of England during the annual Trooping the Color ceremony.

Profile: The Pony

Ponies are not just small horses, but a distinct breed of horse. There are several different breeds of pony, all of which measure less than 14.2 hands tall when they are fully grown. Ponies tend to have a bouncy stride and are naturally more intelligent and curious than horses. Most have a calm temperament and rarely cause trouble for a rider. This is why they have become popular riding horses for both children and beginners.

Working ponies

Due to their size and calm temperament, ponies were well suited for working in factories or even in mines. They could pull heavy loads through narrow spaces and did not panic at the noise or the dark. In 1900, there were 70,000 ponies working in mines in Britain alone, but machinery has since taken their place and nowadays no ponies work in such unpleasant conditions.

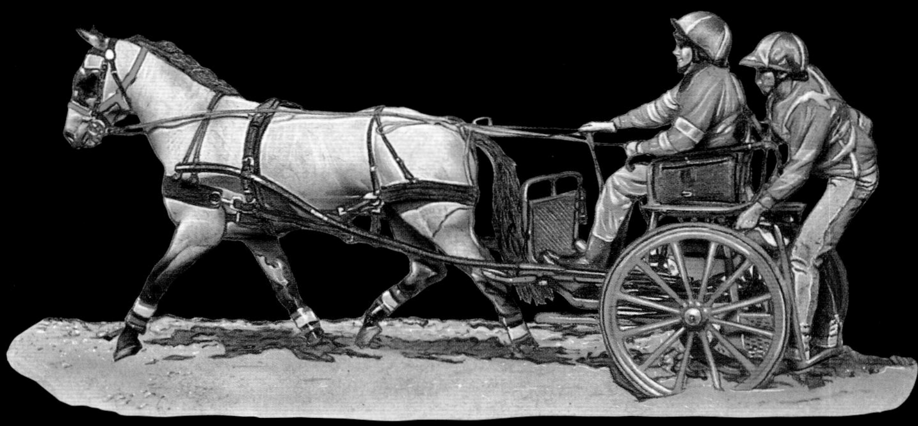

A Connemara pony pulls a two-wheeled trap during a competition.

Pony ID

Origin: various places, mostly in northern Europe
Background: half-wild horses adapted to harsh conditions on mountains or moorland
Height: 14.2 hands or under, often much smaller
Color: wide variety of colors
Uses: riding horses for beginners or children
Characteristics: curious, intelligent, and calm

Harness and riding ponies

Ponies are strong and willing and many of the larger breeds are still used to pull lightweight carriages and carts in some parts of the world. They can cope well in rough conditions and on rugged terrain. For these same reasons, ponies are ideal for cross country racing and hunting.

Ponies and children

Ponies are smaller than horses and have a pleasant temperament. For this reason, they are often used to help children learn how to ride. The cute expressions and large eyes of many ponies make them particularly appealing to young children. Riding ponies is also believed to help people who have been injured in an accident regain strength and feeling in damaged limbs.

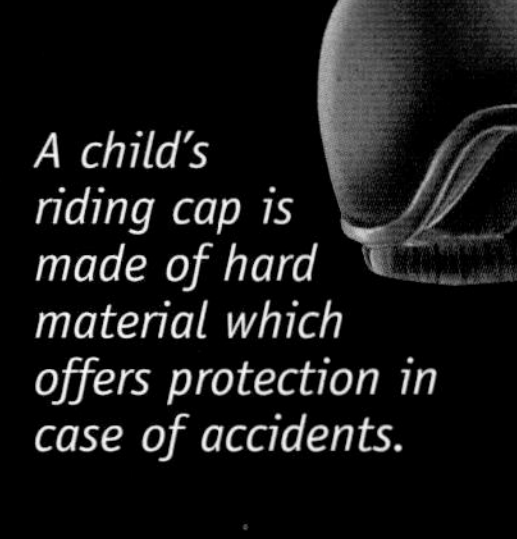

A child's riding cap is made of hard material which offers protection in case of accidents.

Pony clubs

The Pony Club is an organization for child riders with branches in over 30 countries around the world. Pony clubs holds meetings to improve riding skills as well as offering lessons and advice on horse care.

A horse is dangerous at both ends and uncomfortable in the middle.
Ian Fleming (1908–1964), author

THE SHETLAND PONY

The Shetland Pony, sometimes called a Sheltie, comes from the Shetland Islands off northern Scotland. They are small, tough animals that can survive on the poor grass of these cold islands. Shetlands stand less than ten hands tall and are remarkably docile. They are still widely used as riding mounts for children and are now found across the world.

Pegasus and Bellerophon

The ancient Greeks believed in a flying horse called Pegasus, who, according to legend, was born from the blood of the snake-headed monster Medusa. At first Pegasus belonged to the hero Perseus, who gave him to the warrior Bellerophon. When Bellerophon angered Zeus, the king of the gods, the angry god sent a fly to bite Pegasus under the tail so that he would throw Bellerophon. Pegasus then followed Zeus and became a constellation of stars in the night sky.

The Centaur

The Greeks believed in a race of creatures, half men and half horse, who lived far to the north. These creatures, called Centaurs, were said to be wise and clever, but also bad-tempered and violent. It is thought that the stories of Centaurs might be based on the first horse-riding peoples from central Asia, who reached Greece many years earlier.

Mythical Horses

The beauty and grace of horses have captured the imagination of humans for centuries. Their universal appeal can be seen in the way that they have found their way into myths and legends in cultures the world over — as gods or goddesses, or even as powerful and mythical creatures in their own right. Universally portrayed as being brave, strong, and resourceful creatures, horses have always been held in high regard.

Trojan horse

According to Greek legend, the Greeks fought a war lasting ten years with the citizens of Troy, an ancient city in modern Turkey. The Greeks built a large wooden horse, which the Trojans thought was an offering of peace. When the Trojans pulled the wooden horse into their city, the Greek soldiers who had been hiding inside leapt out to open the city gates and killed the Trojans while they slept.

Epona

Epona, the horse goddess, was one of the most important deities of the Celts who lived across western Europe before being conquered by the Romans about 2,000 years ago. She was responsible for the health of horses and was often worshipped by soldiers in Roman cavalry regiments. Epona is usually shown holding a key, a symbol of her control over the entrance to the Otherworld where gods and spirits lived.

Sleipnir

Odin was the great warrior god of the Vikings of northern Europe. This mighty god rode across the world of humans mounted on Sleipnir, a magical horse with eight legs, who was said to be the fastest and most powerful horse of all time. Odin gathered up the spirits of dead warriors and took them to feast with him for eternity at his great hall of Valhalla.

The Unicorn

In the Middle Ages, people in Europe believed that the unicorn lived in lands far to the east. It was supposed to be a horse with a single, spiral horn growing from its forehead. It was said that the unicorn was a wild and ferocious beast, but that it could be tamed by a pure young woman. It is thought that the description of the unicorn may be a garbled account of the rhinoceros.

This French tapestry, woven about 1500, shows the Lady and the Unicorn.

Hindu mythology

Hindus believe that the god Vishnu has appeared on Earth to save mankind on several occasions, each time taking on a different form, or avatar. Hindus believe Vishnu will appear once more at the end of time when he will ride across the Earth mounted on his horse, Kalkhi.

A dancer (right) plays the part of Vishnu's horse, Kalkhi.

Pulling the sun chariot

The Greeks believed that Apollo drove the sun across the skies each day in a chariot pulled by four horses. Apollo was also thought to be the god who caused crops to grow and to ripen as the sun nourished them through the summer. It was believed that he could see into the future and would sometimes reveal what lay in store for humans through an oracle at his temple at Delphi.

A fountain in the shape of the Greek sun god Apollo at Versailles in France (below).

A beautiful carving of the horse god, Ayyanaar (right), from Tamil Nadu in southern India.

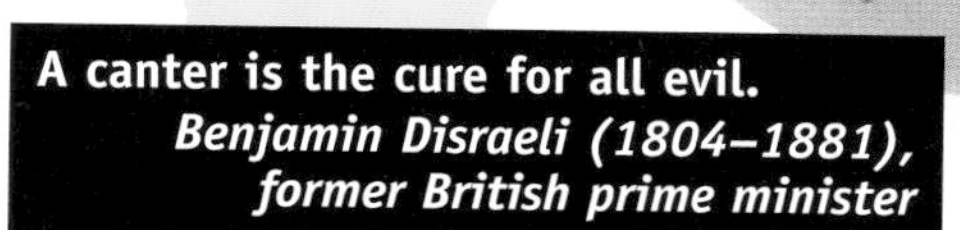

A canter is the cure for all evil.
Benjamin Disraeli (1804–1881), former British prime minister

The Ayyanaar

For centuries in India, carvings of guardian spirits or gods have been erected in villages to protect them from harm. It is believed that the horses keep watch for famine, disease, drought, or enemy warriors. When they see danger, they gallop off to fetch the hero Ayyanaar to protect the village. The horse sculptures are often made of terracotta and may stand up to 19 feet, 8 inches (6 m) tall. Some are over 1,000 years old.

Horse spirits in China

The Chinese have a number of horse gods. Lu Ma is shown as a green horse and is thought to stop quarrels or rumors as well as to bring good luck. Ma Shen is shown as a horse held by a groom. He is said to cure sick horses and is able to kill demons. Ma Wang, or the King of Horses, is the patron god of anyone who works with horses. Ma Wang used to be worshipped by cavalry regiments in the Chinese army.

Ancient races

Over 2,000 years ago, the ancient Greeks and Romans loved watching chariot races. Each chariot was pulled by four horses and hurtled around an oval track with tight corners at each end. The audience sat in banks of seats alongside the race track. The Circus Maximus, the chariot racing stadium in Rome, could hold more than 250,000 spectators.

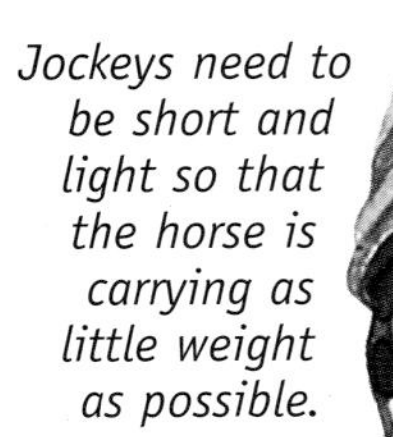

Jockeys need to be short and light so that the horse is carrying as little weight as possible.

Horse Races

People have been racing horses to see who has the fastest one ever since they first learned how to ride. Horse racing today is a large industry that employs a vast number of people. Horses that win races are awarded important trophies and considerable sums of money. People bet money on which horse they think will win a race and some people go to the races just to enjoy the fun. There are many different types of horse racing, each of which is carefully regulated to prevent any form of cheating.

Jockeys

Professional horse riders employed to ride racehorses are called jockeys. A jockey needs to know how to get the best out of a horse in a race, whether he should hold the horse back, or whether he should make the horse run on quickly to gain the lead. Some of the most famous jockeys have ridden hundreds of winners.

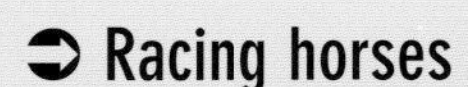

Racing horses

Modern horse racing was started in the 1670s by England's King Charles II (reigned 1660–1685) who put up prizes to be awarded to the winner of a race open to all comers. In the 1770s and 1780s, race meetings began to be held with several races staged on the same course on the same day. These established the pattern of holding fairly short races at set times so that people could place bets on the horses. This way of conducting a program of races at a single venue still continues today.

A medal (above) showing Man-o-War, a racehorse that won more than 90% of the races in which it ran.

Social events

Many race meetings are important social occasions and attract thousands of people. The Royal Ascot meeting is held near Windsor in England each year and draws a rich and fashionable crowd, as does the Prix de l'Arc de Triomphe meeting in Paris, France. The Derby Meeting at Epsom, also in England, attracts thousands of ordinary Londoners. It is held on open hills and, as well as the racing, has a fair and sideshows.

Steeplechasing and flat racing

There are two different types of horse races. Steeplechases are races that include fences (as above). The fences are usually quite small, but in some races they can be over 6 feet, 6 inches (2 m) tall and other jumps include ditches or pools of water. Steeplechasing takes place in the winter. In the summer, horse racing takes place over flat courses, with no obstacles of any kind. This form of racing is known as flat racing. Horses do not generally take part in formal racing until they are three years old.

> **Nobody has ever bet enough on a winning horse.**
> ***American proverb***

Harness racing

Horses pulling carriages have also been raced, as well as horses carrying riders. In North America, this sport has developed as harness racing. Horses are harnessed to a very light, two-wheeled carriage and are raced around oval tracks. The horses are not allowed to gallop or canter, but must trot. Harness racing also takes place in Australia, Russia, and in some countries in Europe.

The trophy given to the winner of the 1999 Kentucky Derby (right) was made of silver with a gold horse and was decorated with 12 emeralds and 350 rubies.

Great races

The horse racing calendar is dominated by great races, known as The Classics, which are held on the same date each year. In April, the British 2,000 Guineas is held at Newmarket. The Kentucky Derby is run in May and the original Derby in June. In September, the Prix de l'Arc de Triomphe is run in France. In November, Australia holds the Melbourne Cup.

Horses in Sports

There are many sports, as well as horse racing, that involve horses. Equestrian events are usually popular wherever they are staged, be it in a village square or in a stadium seating thousands. Some events test the skills of both the horse and rider at performing difficult tasks. Other sports are team games, in which lightning reflexes and superb horse control are essential. They all show off the grace and beauty of the horse.

Classical Riding

The formal, strict rules of classical riding come from the steps and tricks that were needed by a horse ridden into battle in the Middle Ages. During the 17th century, these skills became valued as the mark of a well-trained horse. In the 1730s, the Frenchman François Robichon de la Guerinère brought the various skills together into a formal school of horse training in Paris. This remains the basis of classical riding to this day.

Showjumping

The sport of showjumping, where horses have to jump over decorative artificial jumps, began in Ireland in the 19th century. Each horse and rider take turns to ride over a course of demanding jumps. If the horse knocks down a jump, it acquires four faults and if it refuses a fence, it gets three faults. The horse with the fewest faults is the winner. If two horses have the same number of faults, the winner is the horse that completed the course in the quickest time.

A showjumper (right) clears an obstacle designed to look like a country fence.

Horse trials

The modern horse trial takes place over three days. On the first day, the horses take part in dressage competitions and on the second day in show jumping. The third day is the most demanding, as this is when the cross-country race takes place. Horses are ridden over a series of formidable walls, ditches, fences, and hedges, and are even taken through rivers.

I can make a General in five minutes, but a good horse is hard to replace.
Abraham Lincoln (1809–1865), 16th American president

Dressage

The sport of dressage developed in the 18th century from the training of cavalry horses. The movements are not as formal as in classical riding, but the horse and rider need to work in perfect balance with each other to carry out a series of complicated and difficult steps, such as turning on the spot, walking sideways, changing stride in mid-canter, and so on.

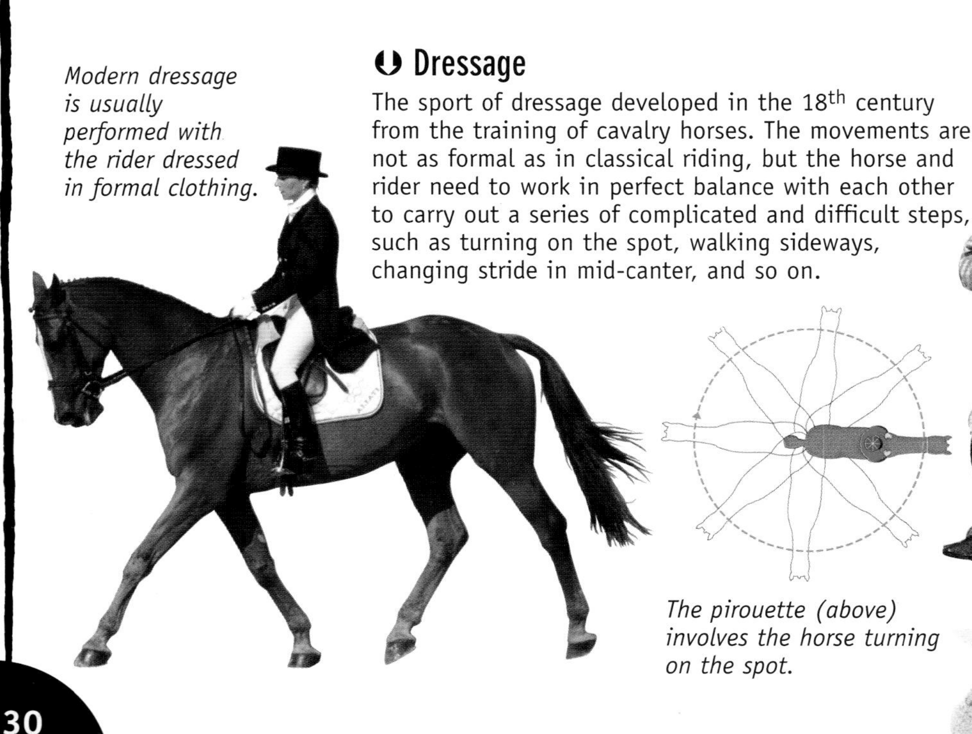

Modern dressage is usually performed with the rider dressed in formal clothing.

The pirouette (above) involves the horse turning on the spot.

➲ Rodeo

The sport of rodeo began in the west of the United States when cowboys competed with each other in skills useful for their work. Events include roping cattle, wagon racing, cattle driving, and riding bucking broncos, or untamed horses. As cattle farming has spread around the world, rodeo has become a popular sport in places as far afield as Australia and Argentina.

Galloping games

Many traditional horse sports demand superb riding skills. In Hungary, the Czikos race involves a rider driving a team of carriage horses while standing on the backs of the rear pair. In Khazakstan, the sport of Oodarysh is a form of wrestling on horseback. In Azerbaijan, Papakh-oyuno is a team game in which players try to steal each other's hats while riding at a gallop.

Polo

The fast, exciting game of polo dates back over 2,000 years to ancient Persia. The opposing teams of four players try to drive a ball into the goal by hitting it with a polo stick. The pitch is 328 yards (300 m) long and 219 yards (200 m) wide and the game is divided into eight seven-minute periods of time called chukkas.

Buzkashi

Whenever the Kirghiz people of Afghanistan hold a celebration, such as a wedding, they organize a game of Buzkashi. The riders divide into two teams, up to 50 riders strong, and fight for control of a dead goat or calf, which may weigh up to 88 pounds (40 kg). The carcass must be carried to a goal and placed on the ground. The players are allowed to use whips and fists in any way they choose in this ferocious sport.

The German artist Franz Marc (1880–1916) painted hundreds of pictures of horses during his career, including this one entitled Little Yellow Horses.

This statue (below) is nearly 2,000 years old and shows the Roman Emperor Marcus Aurelius (180–121 B.C.) mounted on his horse.

Horse Fancy

Horses have featured in art and literature since the very earliest days. Stories about the strength, skill, and intelligence of horses abound in the cultures of ancient Greece, Rome, and Persia. Through the course of time, artists have used the physical attraction of the horse as an inspiration for some of their greatest works.

Equestrian statues

There are many statues of soldiers, politicians, and rulers on horseback. The nobility of the horse adds to the dignity and importance of the person. Casting such large statues from metal is a difficult task. The ancient Romans knew how to do it, but the skill was then lost and remained so until the 16th century. In the 19th century, it became the fashion to show soldiers who had died in battle mounted on rearing horses with both front legs off the ground.

Horses in art

Painters have depicted horses in a variety of roles. This famous portrait of a French cavalry officer by Theodore Gericault (1791–1824) was painted in 1812 and shows him mounted on his Arabian charger. Some artists specialized in horse portraiture. The 18th-century English painter George Stubbs (1724–1806) made an entire career out of painting portraits and writing books about horses and other animals.

The cover of this edition of the classic children's book Black Beauty shows the heroine being nuzzled by his mother.

Horses and children

Horses have been a popular subject for many children's books. Perhaps the most famous of all is *Black Beauty*, written by the English author Anna Sewell in 1877. The story is told by Black Beauty himself as he traces his life from being a foal through to his work as a carriage horse and cavalry charger to his happy retirement.

Right: A horse ornament made of beads by Native Americans. Below: A bronze bridle ornament from ancient Europe.

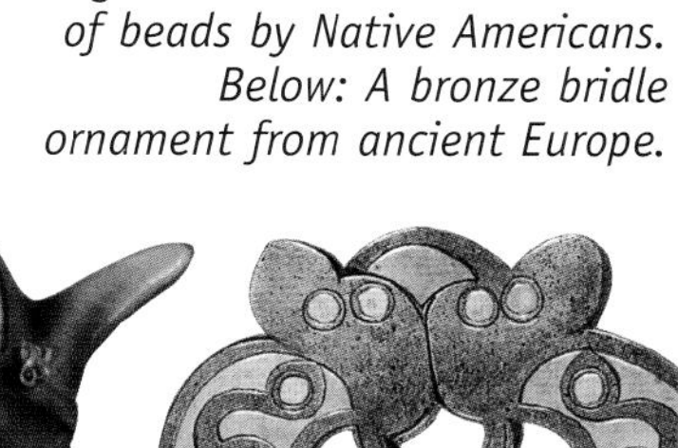

Horse accessories

Proud horse owners have long made their horses look more attractive and important by adding beautiful ornaments to the saddles and bridles. Some of these ornaments included decorated buckles and straps or good luck charms; others were included just to add grandeur to the horse's appearance.

This elaborate carousel horse (right) was carved between 1912 and 1916 in America.

Below: This decorated spur belonged to an American cowboy.

The horse! The horse! The symbol of surging potency and power of movement, of action, in man.
D. H. Lawrence (1885–1930), author

This Mongol saddle (right) is made of carved wood covered in leather.

Left: This poster advertises a spectacular outdoor show of horsemanship staged by the famous hero of America's Wild West, Buffalo Bill (1846–1917). The show toured Europe in the latter part of the 19th century.

Showbiz horses

Horses have played an important role in movies and in shows. The first feature-length movie was a Western called *The Great Train Robbery*, which showed an exciting chase on horseback. The 1944 movie *National Velvet* told the story of a young girl who trains a horse to take part in a major horse race. The more recent animated movie *Spirit — Stallion of the Cimarron*, tells the story of a wild stallion in the American West.

Right: A poster for the film National Velvet.

Studs

Farms where horses are bred are known as studs. The farm usually keeps several male horses (called stallions), and uses them to breed with female horses (called mares), which are brought in from other farms or stables. Only the very best horses should be used for breeding, so that they pass on their superior qualities to the next generation. A good stud farm owner will search for horses with the right temperament for the purpose he prefers, be it racing or country riding. Most studs are found in normal farming areas with plenty of good grazing land. The superior Pompadour Stud in France (left) is based at an old castle.

Draft horses

Large, heavy horses that are used to pull carts or ploughs are called draft horses. They may belong to one of several different breeds. They tend to be large and immensely strong, but can only move fairly slowly.

Horse Breeds

There are dozens of different horse breeds found across the world. Some are only popular in one country or in a small region, but others are more widely spread. Each breed of horse has its own particular characteristic which makes it suitable for specialized work or for a life in a certain type of climate. Some breeds are highly prized — a good specimen of a popular breed can be worth a large amount of money.

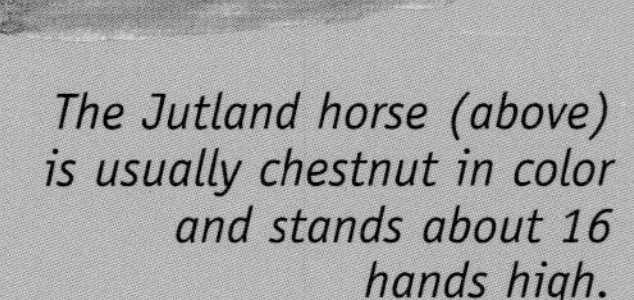

The Jutland horse (above) is usually chestnut in color and stands about 16 hands high.

The Boulonnais (right) comes from northern France and can move more swiftly than most draft horses.

Types of horses

All the breeds are divided into "cold-bloods," meaning larger horses of a docile temperament, or "hot-bloods," horses that are faster and lighter, but which are also more excitable. Some "warm-bloods" are the product of cross-breeding the two types of horse.

A heavy horse used for pulling ploughs or carts is a typical cold-blood breed.

A light horse used for riding is a typical hot-blood breed of horse.

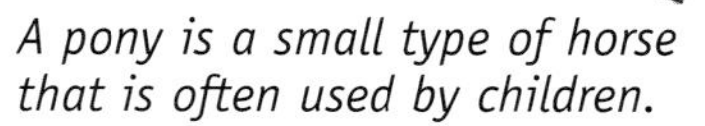

A pony is a small type of horse that is often used by children.

➲ Don

The Don breed comes from around the Don River in southern Russia. It was once used by cavalry regiments because of its powers of endurance and its ability to survive harsh weather. Nowadays it is used more as a riding horse.

➲ Hanoverian

This large, powerful riding horse comes from the city of Hanover in Germany. Hanoverians are easily trained and are agile enough to perform many different tricks or steps, making them ideal for dressage competitions.

Lipizzaner

This race was developed in Austria over 400 years ago by crossing the Andalusian with Berber, Arab, and local horses. Lipizzaners can perform highly formal dressage including spectacular leaps and are famous for their exhibition work at the Spanish Riding School in Vienna.

Andalusian

This breed of riding horse comes from around the city of Jerez in southern Spain. They are known to be particularly intelligent horses and have an elegant, high-stepping gait.

Thoroughbred

This famous racing breed was produced in England by crossing Arabs with native English riding horses. The Thoroughbred is the fastest of all horses, but it has little stamina and is rarely used off the racetrack.

Barb

This North African horse originated in the semi-desert lands of Morocco and Algeria. The Barb often has a long, luxuriant growth of hair on its tail and mane. It is tough and is able to survive on little fodder or water. Although it is popular as a riding horse across North Africa, it is rarely found elsewhere.

Appaloosa

This riding horse from the United States has a beautiful spotted coat. The markings are often just on the rump but can also be spread all over the body. Nimble and with a friendly temperament, the Appaloosa is ideal for riding in hills or on rocky terrain. It is also used as a circus and performing animal, where its agility means it can be trained to do many tricks.

Stabling

Stables provide horses with shelter from the rain or the wind and provide warmth in cold weather. A good stable will have enough room for the horse to move about and there should always be plenty of fresh air and natural light. Each horse is usually kept in a separate stall made of wood, which contains a water bucket and a manger containing food. The floor of the stall should slope so that it is easier to hose down with water. Stables usually include a separate room where the tacks and harnesses are kept, together with a loft for storing food.

Horse droppings and used straw should be cleared regularly.

Feeding

Wild horses eat grass throughout the day and much of the night. Domestic horses can also be fed on grass or on hay (dried grass), but may need supplements to their diet. Oats and barley provide the horse with more energy than grass or hay and should be fed to horses that work hard. Sweet vegetables, such as carrots or turnips, can also be given as a treat. Special feeds with added vitamins, minerals, and proteins can be bought. They are often fed to horses who do specialized work or to those that are suffering from health problems.

Caring for Horses

Domestic horses need to be properly cared for if their owners are to get the best out of them. They should be well fed, given shelter in harsh weather, and any sickness should be treated promptly. A horse's equipment should be well made and well cared for and should always be the right size and shape to fit the horse properly.

Grooming

Horses need to be groomed to remove dirt and insects from their coat. The horse should be brushed thoroughly all over, then the hooves should be picked clean of stones and coated with protective hoof oil. Finally, the horse's eyes, nostrils, and mouth should be checked for signs of infection or insects.

A Mongol warrior grooms his horse.

Caring in the open air

Some breeds of horse are hardy enough to endure life outdoors all year round, but others need to be taken indoors in the winter. Many horses can be kept outside in cold weather if they wear a rug. The rug is a carefully fitted waterproof sheet lined with warm woollen cloth. In cold weather, extra woollen rugs can be put underneath to provide the horse with extra warmth.

➲ Saddlery

There are many different types of saddles for riding horses. A normal riding saddle is made of smooth leather fitted over a solid frame. It spreads the weight of the rider and acts as a support for the stirrups. Cowboys use a saddle with a solid knob round on which they can wind their ropes. Ladies wearing long dresses cannot ride astride a horse, so they have a sidesaddle and keep both legs on one side of the animal.

Left: A 19th-century sidesaddle.

Half the failures of this world arise from pulling in one's horse as he is leaping.
Augustus Hare (1834–1903), author

Left: Horse hoof protector.

Ferriery

Owners should take constant care to protect a horse's hooves. Metal shoes protect them from excessive wear on roads and help the animal cope with the extra weight of a rider. The hooves should be checked regularly for stones or infections. If a hoof becomes cracked or injured, it can be protected by a leather shoe, under which, in severe cases, hot poultices or packs can be placed.

A horse o-ring.

A horseshoe.

Training and riding

Training a horse is a lengthy and time-consuming business. A young horse should be handled by humans from the day it is born, but formal training need not begin for some weeks. At first, the foal is led by a halter, and then it is introduced to being tied up and shoed. At about the age of two years, the horse is broken in, meaning it is trained to carry a rider and to respond to movement on the reins. Specialized training of working horses, such as police horses or cattle horses, can take many months of patient work by the trainer.

The metal bit in the horse's mouth transfers instructions from the rider, which are given by pulling on the reins.

A rider mounted on a standard European riding saddle.

The girth strap passes under the horse's belly and secures the saddle.

The saddle cloth stops the leather saddle from rubbing against the horse's skin.

Index